WARM-UP EXERCISE

As important as the skills discussed in Units II and III may be to an engineer, to create complicated machines and structures, he must have something else: theory and practical know-how. The next three learning segments offer a sampling of the knowledge that an engineer must have and some generalizations about it.

If you have trouble thinking of answers to these questions, don't waste time on this exercise. On the other hand, if you think of answers fairly quickly, you might want to re-evaluate your answers after you finish this unit.

1. In nearly all control systems, there is some movement in the value of the controlled variable around the desired value. Can you think of two objectives that would have to be traded off as one tried to optimize the amount of this movement?

____

____

____

____

2. Scientific and engineering activities differ in many ways. Can you think of two ways in which they differ?

____

____

____

____

3. Computers are important to engineers for many reasons. They not only help engineers in the process of solving problems, but, in many cases they also become part of the solution. Can you think of two ways that engineers use computers in the process of solving design problems?

____

____

____

____

4. Can you think of two reasons that an engineer might find for using a computer as a part of a design solution?

____

____

____

____

Learning Segment 10
CONTROL SYSTEMS

LEARNING OBJECTIVES

- Describe the general nature and significance of a feedback control system.

- Recognize the widespread presence of feedback control systems in nature and society as well as in engineering.

- Know the meaning of such terms as *sensor, processor, effector, closed-loop system, oscillation, insensitive zone,* and *response thresholds*.

INTRODUCTION

You undoubtedly know that engineering education includes subjects such as chemistry and physics and, in fact, you may know a fair amount about these subjects. Perhaps you have also been aware of the importance of such subjects as feedback control systems.

Besides offering further insight into engineering, exploration of control systems provides some general knowledge of unusual importance. Control systems have widespread significance in social, political, biological, and economic systems, as well as in engineered systems. What follows is an introduction to control systems in general and then an elaboration on how this knowledge is employed by engineers.

It is impossible to aim your automobile down the highway, take your hands off the wheel, and get very far without something happening. This is so even on a straight stretch, mainly because there are crosswinds, slopes, and bumps that force the car off course. What is needed is a means of *detecting* drift of the car toward either side and a way of *correcting* for that drift as soon as it is detected. The means for accomplishing this are familiar: your eyes enable you to detect the need for corrective action; your brain and especially your reflexes enable you to decide what remedial action is called for; your arms and the steering mechanism are employed to implement those decisions. These parts of you and the automobile constitute what engineers call a feedback control system, the general form of which is described in Figure 1. (The correct term is *feedback control system,* since there are control systems that do not involve feedback. However, technical people tend to omit the descriptor and speak of a control system, feedback implied. Furthermore, many persons refer to the same thing as a feedback system, control implied.)

MODERN ENGINEERING
Knowledge Employed by Engineers

Edward V. Krick

Lafayette College
Easton, Pennsylvania

Wiley Professional Development Programs

Advisory Editor
Steven C. Wheelwright
Harvard Business School

John Wiley & Sons, Inc.
New York • London • Sydney • Toronto

Library of Congress Catalogue Card Number: 76-10038

ISBN 0-471-01702-7

Printed in the United States of America.

10 9 8 7 6 5 4 3 2 1

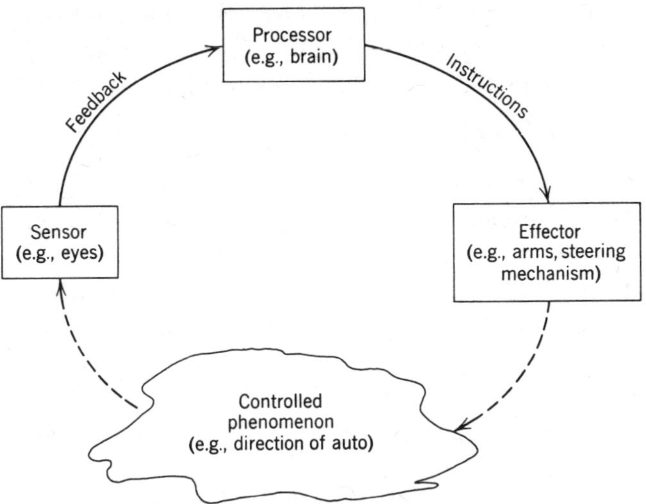

Figure 1. These are the basic components of a feedback control system. A sensor obtains feedback (i.e., a "status report" on what is happening). A processor (decision maker) compares what the sensor reports is actually happening with what is intended (the goal) and decides if and what kind of remedial action is needed. An effector (corrector) executes the corrective action. This cycle of events repeats. You know that as you steer your auto, after you correct your course, you observe the results and if necessary make another correction, and observe the results of it. These activities are usually continuous and simultaneous, not discrete and sequential as described here.

THINK IT THROUGH

There are many activities that you perform every day that involve the functioning of a feedback control system. Name a few.

____

____

____

____

Some possible responses to this exercise appear in the following paragraph.

A feedback control system enables you to ride a bicycle, catch a ball, learn any motor skill, and keep your body temperature close to constant. Even the relatively simple act of reaching to the top corner of this page before turning it involves a rather elaborate control system. The human body contains a remarkable array of such systems, with myriad external and internal sensors, a multilevel decision-making hierarchy, numerous interconnections, and a complex system of glands and muscles serving as effectors. And so it is with all living organisms; recall that even an amoeba can detect and respond to certain changes in its environment.

There are several characteristics of feedback control systems worthy of special mention. One is the closed loop, which is apparent in Figure 1. You can understand why the term *closed loop* is often used in referring to such systems. By "closing the loop," effect is coupled with cause, so that the cause-effect relationship is now one of *interdependence*.

Of course, it is feedback that closes the loop. Feedback is information characterizing the *actual* situation, which the processor compares with the *intended* state of affairs. The discrepancy between intended and actual (the error, in other words) becomes the basis of corrective action.

Another feature worth noting is the role of information throughout a feedback control system. The system must be given a desired condition (intended state of affairs, goal, set point). This is information, so is feedback, and so are instructions transmitted to the effector. The processor converts information from one form to another; the effector converts information into action; the sensor converts action into information. No wonder information is sometimes called the lifeblood of a control system.

We need a feedback system in this learning program. If we were speaking face to face, we could get feedback through facial expressions, questions, and comments to learn if the message is received and understood (the goal). On the basis of the feedback, we could decide whether repeating or restating or exemplification are necessary to achieve the goal. Feedback facilitates learning (and, in the case of motor skills, is essential). Picture the blindfolded dart thrower; he throws darts indefinitely and shows no improvement as long as he knows nothing about the effectiveness of his efforts. Exercises such as the following take the place of face-to-face feedback.

TECHNICAL TERMS

To increase your understanding of the terms we have been using, write a brief definition for each term. Then, help make our learning feedback and control system work properly by checking your definition against that given after the exercise and make any needed adjustments on the basis of this feedback. You may want to check each definition as soon as you write it in order to see how this immediate feedback might affect the next definition.

1. Sensor

____

____

____

2. Processor

---

3. Effector

---

4. Closed loop

---

5. Feedback control system

---

1. The sensor in a feedback control system is the element that detects the actual state of affairs regarding the phenomenon that is to be controlled. The sensor passes the information detected on to the processor, or decision maker. Sensors include eyes, ears, kinesthetic (muscular) sensations, or any mechanical monitoring device.

2. The processor in a feedback control system is the decision maker that compares the actual state of affairs (as reported by the sensor) with the desired or intended state of affairs and decides or determines what, if any, corrective action is required. Processors include the human brain, a computer processing unit properly programmed, an animal's central nervous system, and so on.

3. The effector in a feedback control system is the means used to carry out, or effect, the corrective action to modify the actual state of affairs of the controlled phenomenon in a way determined by the processor. Effectors include body muscles and limbs, a steering or other stabilizing mechanism, and so on.

4. Closed loop in a feedback control system refers to the fact that there is a flow of information from the controlled state of affairs to the sensor to the processor through the effector to the controlled phenomenon and back to the sensor to restart the feedback control cycle over again.

5. A feedback control system is a closed-loop process that senses the actual state of affairs in a phenomenon to be controlled, determines what controlling or corrective action is to be taken, effects the action and continues to monitor the controlled phenomenon for any further needed changes.

AUTOMATIC CONTROL

In the control systems mentioned so far, humans predominate. But this is certainly not necessary; man has learned how to create devices that will perform control functions, many of which require no direct intervention by humans. In general, engineers are responsible for designing these *automatic control* systems.

A familiar example of automatic control is the system that maintains the temperature at a comfortable level in modern buildings. The thermostat performs the sensing and processing functions, instructing the furnace (the effector) when to cut in and out in order to keep room temperature (the controlled phenomenon) at the specified level. The human's only role is to prescribe the desired level by setting the thermostat dial.

A classic example of automatic control is the flyball governor applied by James Watt to control the speed of a steam engine (Figure 2). This ingenious device made it unnecessary for a man to stand by to adjust the steam input, as the load on the engine changed.

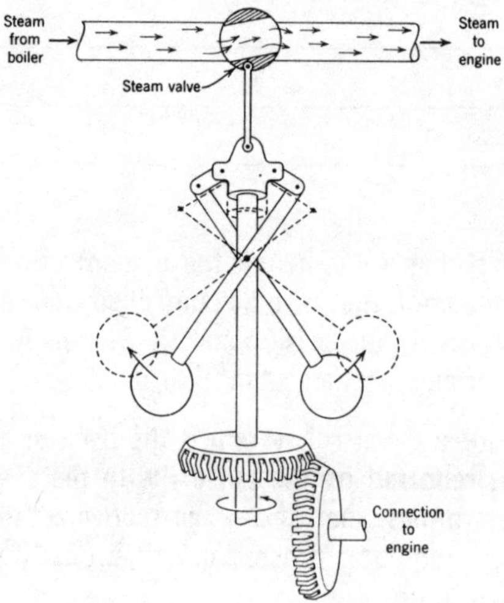

Figure 2. The flyball governor is connected mechanically to the output shaft of a steam engine so that the ball mechanism rotates at the speed of the engine. If the load on the engine decreases, speed will tend to increase which, through centrifugal action, forces the balls outward. Through the linkage, this will proportionately close off the steam supply to the engine. If the engine tends to lose speed, the mechanism increases the steam supply accordingly. Therefore, the flyball governor maintains engine speed at a preset value without human intervention. This invention is significant in several respects. It is remarkable if for no other reason than it was so advanced for its time (the 1780s). Furthermore, it is a classic illustration of the elegant solution. Finally, it is widely recognized as an outstanding example of what engineers can do without the benefit of theory. The mathematical theory of the behavior of this governor did not appear until 1868.

Recall the machine for manufacturing reed switches described in Unit I. A sophisticated feedback control system is at work. Finished switches pass through a series of instruments that perform critical measurements and tests, not only to eliminate unsatisfactory switches but to enable the production machine to correct the cause of error. If the machine should start producing switches with oversize gaps between reeds, for instance, it automatically detects this and adjusts the gap-setting mechanism appropriately. This machine is a prime example of automation; so are the thermostat system and the ship positioner in Figure 3.

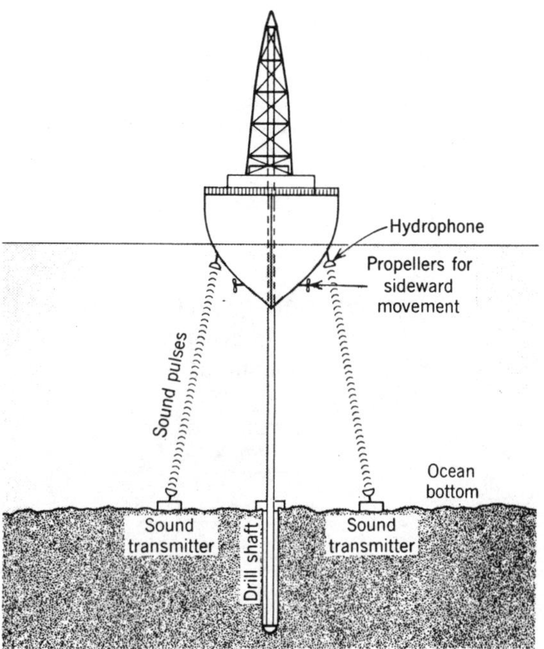

Figure 3. A drillship is an ocean-going vessel equipped with the familiar drilling rig and other special apparatus that enables it to drill into the ocean bottom in deep water. When drilling, such a vessel cannot drift very far from the desired position directly above the drill hole. The automatic control system enables the ship to maintain its position in the face of winds and currents. Beacons transmit sound waves from known positions. Theses are sensed by hydrophones and, from their phase relationship, the computer can calculate drift. On the basis of this, the thrust propellers fore and aft are actuated to return the ship to the desired position.

LET'S BE SURE ABOUT THIS

What is the essential difference between automatic control systems and feedback-control systems in which there is a human functioning as the processor?

____

____

In the diagram, fill in the specific items from the automatic control system for drilling ships shown in Figure 3.

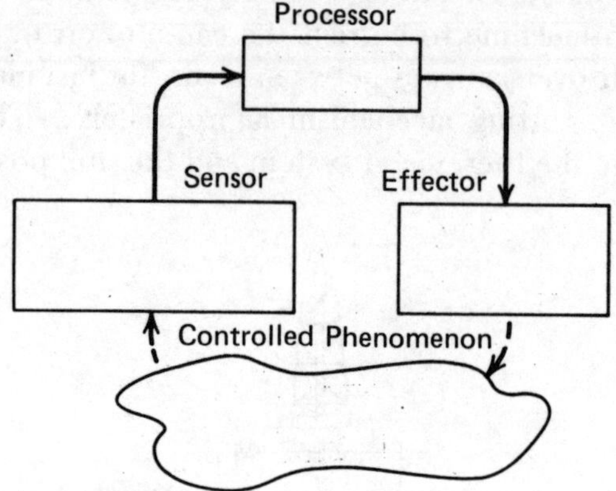

The essential difference between automatic control systems and others in which the processor is a human being is that the human being can walk away from the automatic control system and it will continue to function properly.

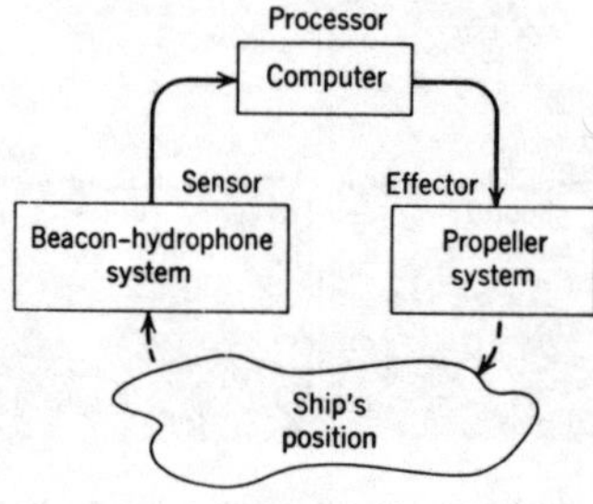

THINK IT THROUGH

A human can walk away from an automatic control system and it will continue to function properly. But this not also true of a number of automatic machines that are not automatic feedback control systems? What is the difference?

____

____

____

____

An answer to this exercise appears in the following paragraph.

The same is possible with the automatic washer and the familiar traffic signal, but not for the same reason. These mechanisms are programmed by cams that force the machines to follow a cycle of actions. As long as these machines are functioning properly, they will continue plodding, oblivious of the appropriateness of their actions; the clothes can be clean or filthy or there can be no clothes in the machine; the traffic can be heavy in one direction or the other, yet the signal cycles on. In contrast, the thermostat, ship positioner, and switch-making machine require no human intervention because they have *feedback* control systems that enable them to adjust their behavior in response to changing conditions. These are often referred to as *self-regulating* or *self-correcting*.

YOU DESIGN IT

What do you think keeps the blades of a windmill facing into the wind? You probably never gave the matter any thought. Please do. In fact, do a little investigating in the library before accepting the brief answer presented here.

Explanation of what keeps a windmill facing the wind:

____

____

____

____

____

Diagram:

On smaller windmills, a simple vane perpendicular to the plane of the blades does the trick. As soon as the wind changes direction, unequal wind pressure on the vane automatically makes the correction (but not without some oscillation).

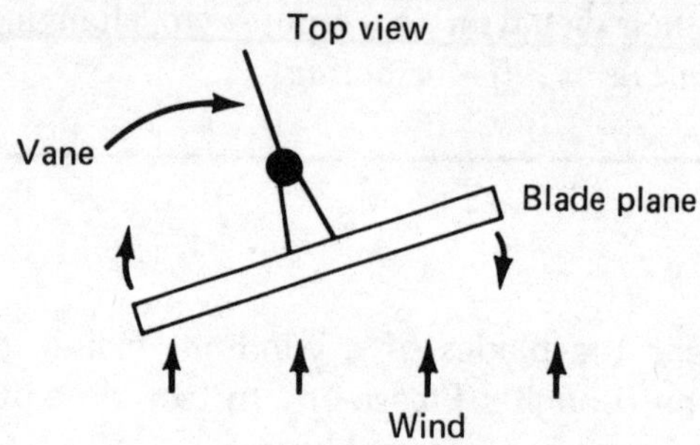

On larger windmills, such as those found in the Netherlands, this elegant solution is not up to the job; those windmills are large and heavy. One method used by Europeans is a small windmill placed behind and at right angles to the main blades. A change in wind direction causes the minor blades to turn, and through a gear train, they restore the major blades to the face-the-wind position.

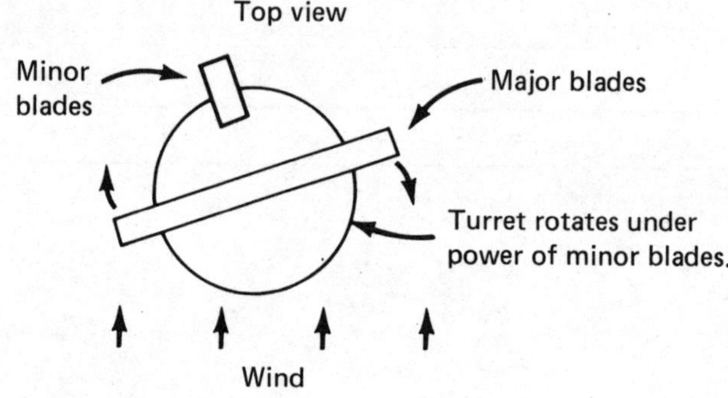

DESIGN OF FEEDBACK CONTROL SYSTEMS

There are feedback control systems in most engineering creations—automobiles, power-generating stations, dams, elevators, electric coffee-makers, water supply systems, and steel-making furnaces, to name a few. So the design of such systems is a rather common activity in engineering. Ordinarily, it all begins with a specific need for control (e.g., a system to maintain airplane cabin pressure within acceptable limits). To synthesize a system that satisfies this need, engineers must know a lot about control systems, in particular such matters as response characteristics, oscillation, sampling, sensitivity, stability, nonlinearity, noise, and a variety of other mysterious-sounding topics.

To give you some notion of what is involved in the design of control systems and what engineers must know in order to design them, here is an elaboration on one of the many matters that must be considered: response thresholds (action limits), chosen primarily because no special background is needed to understand what is involved and because it is relevant to control systems wherever they occur.

You have probably noticed that even in a room with a thermostat system, the temperature fluctuates, reaching a low just as the heater cuts in and a high just after it goes off, as graphed in Figure 4. This controller has an upper limit and a lower limit, and only when the room temperature reaches one of these limits does the system respond. In effect, the thermostat has been instructed not to bother the furnace until the temperature changes enough to make startup worthwhile. The thermostat obliges; it "calls" to the furnace for heat only when room temperature falls 3 degrees below the setting on its dial. When heat is called for, room temperature is allowed to reach 3 degrees above the set value before the furnace is signaled to quit.

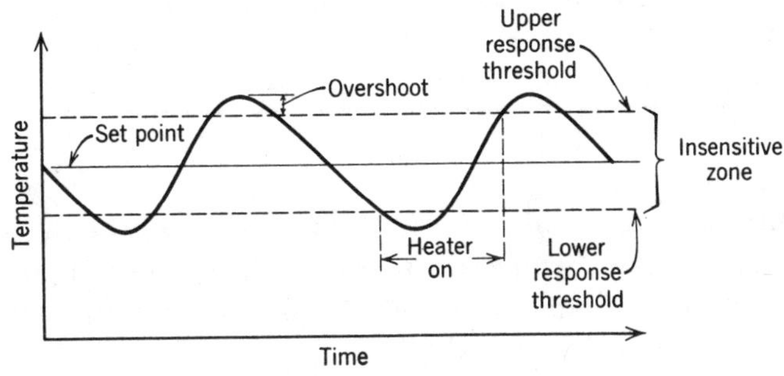

Figure 4.

Response thresholds of a control system bound an *insensitive zone*. Within this zone, the controlled phenomenon can fluctuate without the effector knowing or caring.

TECHNICAL TERMS

Write a brief definition of the following term.

Insensitive zone ____

____

____

____

____

An insensitive zone, in a feedback control system, is the range of values between an upper and a lower limit within which the system ignores any change in the phenomenon to be controlled.

THINK IT THROUGH

Why are insensitive zones built into feedback control systems?

An answer to this exercise appears in the following paragraph.

In many cases, it is infeasible to do otherwise. Take the thermostat system; without an insensitive zone surrounding the set temperature, the controller would go berserk (so would the furnace) going on and off in rapid succession as the room temperature oscillated fractions of a degree above and below the set value.

Here is another use of response thresholds. It is possible to attach a tiny sensor to a critically ill patient to pick up his heart beat. These pulses are transmitted to a monitoring unit at the nurse's desk (Figure 5). That unit sounds an alarm if the patient's pulse rate goes above or below preset limits (right). It makes sense to use such limits when the phenomenon being controlled is subject to a certain amount of natural variation that is no cause for alarm.

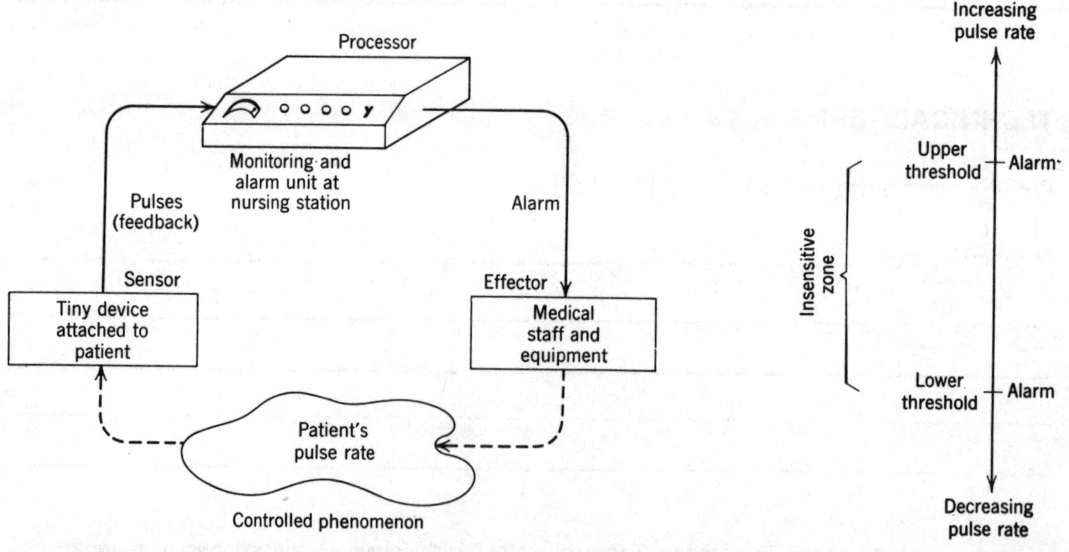

Figure 5. A feedback control system for maintaining a patient's pulse rate within acceptable limits.

MEMORY JOGGER

What process (studied in Unit III) would an engineer go through to determine the most appropriate maximum and minimum limits for an insensitive zone in a feedback control system?

The answer to this exercise appears in the following paragraph.

The manner in which an engineer arrives at the best settings for such limits is an excellent study in optimization. In referring to Figure 6, picture an engineer designing a system for controlling water level behind a dam. It is infeasible to control that level to an exact value; a certain insensitive zone is natural for this situation; but how wide should that zone be? As the zone is widened, there is an increase in the penalties and inconveniences associated with fluctuations in lake level (curve *a*). As the zone is reduced, more expensive control equipment is necessary, and the cost of operating and maintaining the spill way gates increases because they must respond more frequently (curve *b*). Because of these conflicting criteria, there is an optimum zone width, determined by the minimum point of curve *c*.

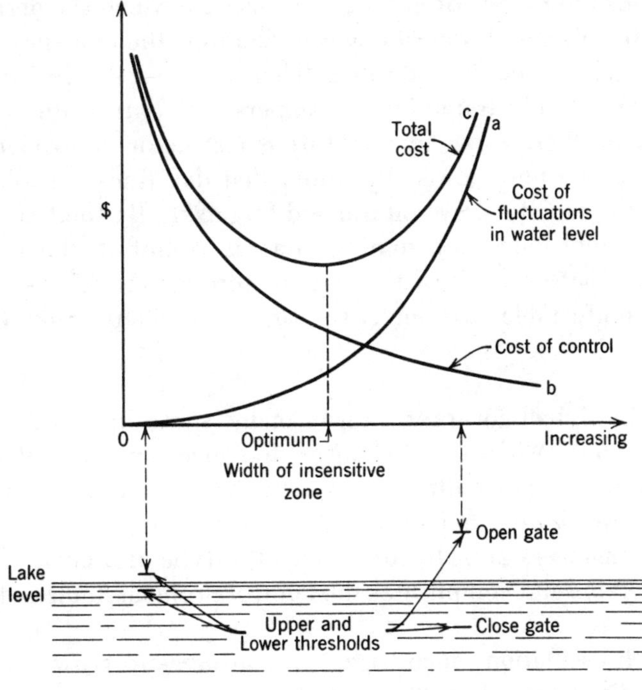

Figure 6. An illustration of optimization in design, in this instance, the design of a system for controlling water level behind a dam.

THINK IT THROUGH

What difficult trade-off problem are you likely to encounter in setting upper and lower thresholds for the alarm system of the pulse-rate monitor outlined in Figure 5?

An answer to this exercise appears in the following paragraph.

If you try to determine the optimum thresholds for the pulse-rate monitor described by Figure 5, you again run headlong into the life-versus-dollars dilemma. The trade-off is between staffing cost and risk to the patient. As you tighten the insensitive zone, you increase the number of false alarms; as you widen it, you increase the probability that a patient requiring emergency attention will not be discovered until too late.

The setting of thresholds is only one of many problems with which the designer of a control system must cope. Probably the thorniest one is oscillation—the tendency of the controlled phenomenon to vibrate about the desired value. There are several types of oscillation. One is continuous, which is nicely illustrated by the temperature control situation in a classroom that we use. There is no thermostat, temperature is instructor-controlled and, usually, it goes like this. The first instructor to use the room that day finds it cool when he arrives. He immediately turns the radiator valve on full and forgets it. By the time the next instructor enters the room it is unbearably hot, so his first act is to shut off the heat. Of course, he doesn't turn it on again before he leaves. And so, throughout the day, the temperature oscillates between uncomfortable extremes. Continuous oscillation like this is commonly referred to as *hunting*.

You don't think that the flyball governor returns engine speed immediately and exactly to the selected speed, do you? When speed changes, the governor brings it back toward the selected speed but, because of momentum, it overshoots. So it must swing in the opposite direction and, again, it overshoots, and so on. But, because of friction, this oscillation subsides—dampens out, as engineers say. In some cases this type of oscillation is inconsequential; in others it is intolerable and complicates the control problem immensely.

There are cases in which oscillation, once triggered, continues to grow. An unstable situation like this could be disastrous. It will probably happen to you when you try to walk on a railroad track. In that case it will probably be of no great consequence, but a hovering VTOL that develops this type of oscillation is in serious trouble.

TECHNICAL TERMS

Write a brief definition of each term.

1. Oscillation ____

____

____

2. Hunting ____

____

____

1. Oscillation is the tendency of a controlled variable to vibrate or regularly increase and decrease around a desired value in a feedback control system.

2. Continuous oscillation around a desired value is called hunting.

THINK IT THROUGH

Briefly explain what causes oscillation in a feedback control system.

____

____

____

____

____

An answer to this exercise appears in the following paragraph.

Ordinarily, oscillation of a control system is the result of delayed response (i.e., delay between the time a change occurs in the controlled phenomenon and the time the system responds). This *lag* causes overshooting—back and forth over the desired level. Lag may arise anywhere in a control system; for example, the feedback may be delayed or the effector may be slow in responding. Certainly a furnace does not produce heat the instant it is triggered, nor does it cease doing so as the power is cut off. Such sluggishness and the resulting oscillation is often apparent in economic and political systems. All control systems suffer from some lag and, therefore, are pone to oscillate to *some* extent. For the engineer it's a matter of keeping that oscillation at an optimum level.

YOU FIGURE IT OUT

You have just read: "For the engineer it's a matter of keeping that oscillation at an *optimum* level." Why optimum rather than minimum?

Reducing the oscillation of a system will cost something; eliminating it (if this is possible) would send the cost out of sight. This is one more illustration of a generality technical people have long since recognized: perfection is uneconomical to achieve, even to approach.

SELF QUIZ

1. Briefly describe a feedback control system.

2. Explain why feedback control systems are of wide significance not only in engineering, but also in many other areas. Give a few examples to indicate this widespread importance.

3. Define the term *sensor*.

4. Define the term *processor*.

____

____

5. Define the term *effector*.

____

____

6. Define the term *insensitive zone*.

____

____

1. A feedback control system is a cycle in which the actual state of affairs in a phenomenon to be controlled is sensed and reported to a decision maker that compares the actual state with the desired state and determines a course of action to modify the controlled phenomenon. The sensor continues to monitor the controlled phenomenon and reports any changes to the decision-making element to start the entire cycle over again.

2. Feedback control systems appear in a wide variety of natural ways in our environment (for example, maintaining the stability of populations of a wide variety of animal species), in our governments and society (checks and balances between branches of government and community pressures on antisocial behavior, for example), as well as in our machines and technological processes.

3. The sensor in a feedback control system is the element that detects the actual state of affairs regarding the phenomenon that is to be controlled and passes the information detected on to the processor, or decision maker.

4. The processor in the feedback control system is the decision maker that compares the actual state of affairs with the desired or intended state and decides what action to take.

5. The effector in a feedback control system is the means used to carry out the corrective action to modify the controlled phenomenon.

6. An insensitive zone in a feedback control system is the range of values between a minimum threshold and a maximum threshold within which the system ignores changes in the controlled phenomenon.

Learning Segment 11
DIGITAL COMPUTERS

LEARNING OBJECTIVES

- Understand the general nature and operation of a computer.

- Explain the meaning of such terms as *data, input, output, memory, program, processing unit, branching, library function, filekeeping function,* and *time sharing.*

- Know and be able to explain the four major computer applications: 1) storing and retrieving information efficiently, 2) processing information efficiently, 3) handling information at speeds that no other means can accomplish, and 4) keeping track of amounts of information that exceed the capacity of other means and acting on that information at speeds that exceed the capacity of other means.

- Recognize the opportunity for a beneficial application of computers.

INTRODUCTION

To receive the full benefit of this learning segment, you do not have to know the details of computer operation. The emphasis here is on applications. However, it will be helpful if you know a few general facts about these machines, beginning with Figures 7 and 8.

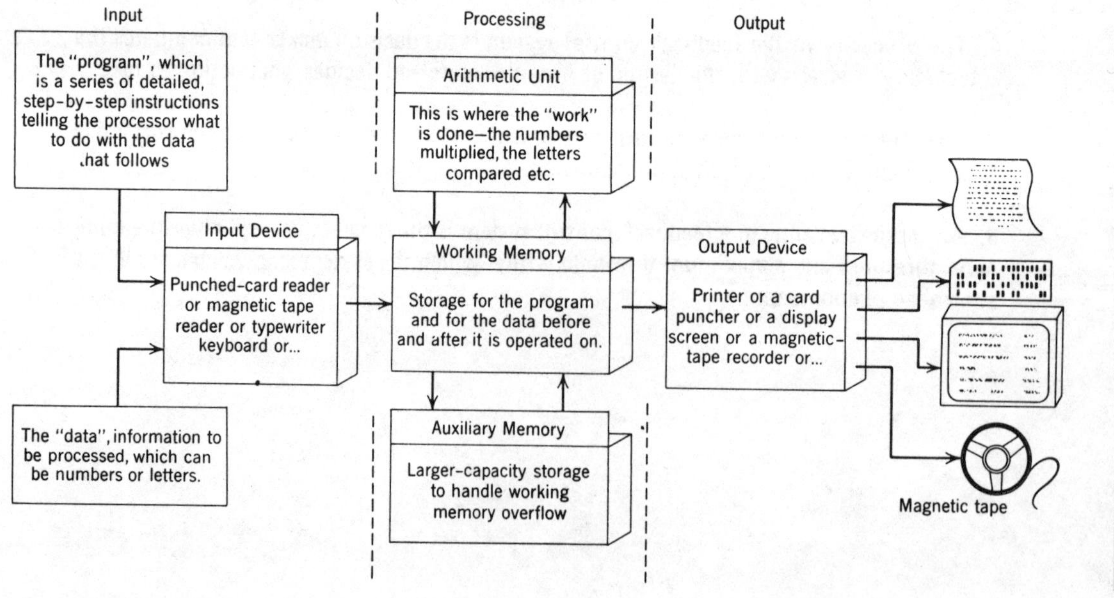

Figure 7. A simplified overall view of a computer system. The processor's "working memory" serves much the same purpose your mind serves in memorizing computational procedures and paper serves when you are doing arithmetic by the pencil and paper method. Although this memory is incredibly fast, it has a relatively limited capacity and, furthermore, it is "cleared" after each job is complete. So, when more capacity is needed or when the user wishes to preserve the memory's contents, an auxilliary storage device like magnetic tape or a magnetic disk augments the working memory. Auxiliary storages are generally slower, but they do have virtually unlimited capacities for letters and numbers—billions of them.

Figure 8. Examples of the computer equipment ("hardware") identified in Figure 7. Not shown is a card reader-punch (a machine that senses the information punched in cards or punches information into them), a paper-tape reader-punch, or a cathode-ray-tube terminal (CRT). A CRT terminal is a close relative to a television receiver in appearance and operation.

LET'S BE SURE ABOUT THIS

1. What two inputs are essential for a computer operation, regardless of the device used to accomplish the input?

____

____

____

____

2. List two functions that must be present in the central processing unit of a computer operation for it to complete its processing activities.

____

____

3. List four forms in which the output of a computer operation might be delivered.

____

____

1. Two essential inputs for a computer operation are data and a program that tells the processor what to do with the data.

2. The central processing unit of a computer operation must have an operating arithmetic function and a working memory that may be augmented by an auxiliary memory. (If you have some knowledge of computers, you may have included control function, too.)

3. The output of a computer operation might be delivered in the forms of a magnetic tape, punched cards, an electronic display screen, or a printed form.

When a computer is used, the sequence of events is generally as described in the following paragraphs.

1. A program of instructions is prepared (or, since programs for many purposes are already prepared, it may be simply a matter of selecting one). These instructions tell the computer what to do with the numbers and letters subsequently given to it, step by step (e.g., "take the number stored in memory location A and add it to the number in memory location B, place the sum in memory location C"). The program even includes details such as where to print the results on the paper. Incidentally, programs are written in terms of "unknowns," like the X's and Y's of algebra. However, in the program you can use words for your unknowns, as illustrated by this excerpt from a program for computing the take-home wage of an hourly employee.

$$\text{WAGE} = (\text{HOURS} \times \text{RATE}) - \text{HOSP} - \text{INSUR} - \text{TAX}$$

Each week, when employee earnings are to be calculated, the same program is entered into the computer and stored in the working memory.

2. Next, the data to be processed are entered into the computer and stored in another part of its memory. In the payroll example, the data consists of the identification number, hourly pay, number of hours worked, and so forth, for every employee. If the number of employees is large, not all of these data will fit into the computer's memory at one time; they will be entered in reasonablesized batches.

3. Now that the data *and* instructions are in the machine's memory, the processor begins operating on that data according to those instructions. In the payroll example, the same calculations must be performed for every employee, so the processor simply recycles and repeats the same series of instructions for each employee.

4. Eventually the results of the computer's operations are communicated to humans by printing them on paper or by other means. In the payroll example, the machine is instructed to print and punch the results on cards, which become paychecks.

LET'S BE SURE ABOUT THIS

List five essential events that must occur in order for a computer to complete a "job" successfully.

1.______________________________

2.______________________________

3.______________________________

4.______________________________

5.______________________________

1. Prepare or select a program.

2. Enter the program into the computer's working memory.

3. Communicate the data to the computer's working memory.

4. The arithmetic unit processes the data according to instructions provided by the program.

5. The results of the processing are "outputted" or communicated to the appropriate person or device awaiting the results.

A digital computer can accept, transfer, and manipulate numbers *and* letters, including letters combined into words and groups of words. It is capable of performing all arithmetic operations. One especially significant capability of a digital computer that you may not know about is its ability to compare numbers *or* letters and follow different series of instructions, depending on the results of the comparison. Suppose a computer is being used to keep driver and auto records for a state. When an application is received for transfer of ownership, part of the computer's task is to check if the auto has been reported stolen. It does so by comparing the auto serial number (e.g., AXB1372594) with each serial number in a list of serial numbers of cars reported stolen, stored in memory. When it compares AXB1372594 with a number on this list, if the two do not match, the computer goes on to the next number (Figure 9).

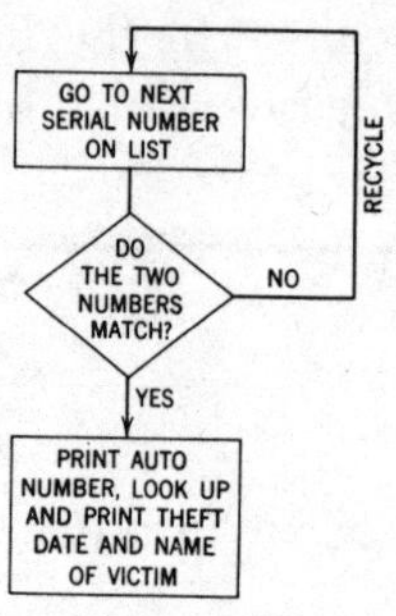

Figure 9.

As long as it finds no match, it continues through the list. If the two match, the computer follows an alternate series of instructions that cause it to print out

AXB1372594 REPORTED STOLEN JAN 6, 1976 BY
JOHN REES 721 CENTER STREET BIRDSVILLE MICH

This important feature, called *branching*, gives the computer more flexibility than you may have given it credit for. There may be thousands of such branches in a long program, so that the computer rarely repeats the same sequence of instructions. Among other things, the computer's branching capability enables it to handle exceptions, e.g., an employee who worked more than the usual 40 hours (Figure 10). You may recall seeing such a diagram earlier explaining how a computer can execute the Monte Carlo process.

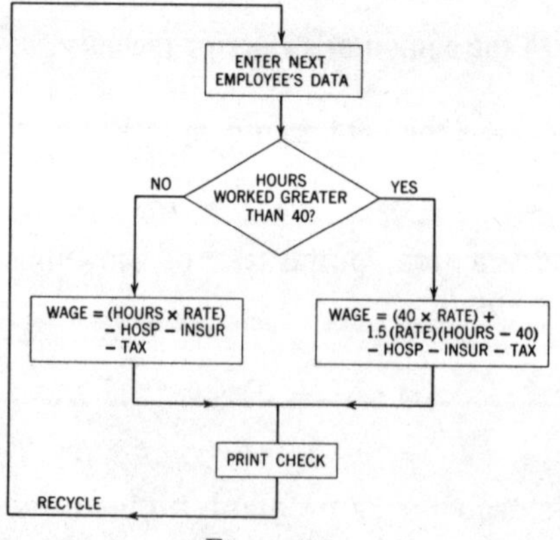

Figure 10.

Of course, there are many variations in computer equipment and procedures for using it but, on the basis of this background, you can develop a general appreciation of what digital computers do.

TECHNICAL TERMS

Write a brief definition for each term.

1. Program ______________________________

____

____

2. Data ______________________________

____

____

3. Working memory ______________________________

____

____

4. Branching ______________________________

____

____

1. A program is a set of instructions that tells a computer what to do.

2. Data is information that is operated upon by the central processing unit of the computer.

3. A computer's working memory is needed to store the computer program and the data that is to be operated upon by the computer.

4. Branching is the computer's ability to compare numbers or letters and follow different series of instructions, depending on the results of the comparison.

AN OVERVIEW OF THE ROLE OF COMPUTERS IN ENGINEERING

Ability to program a computer (i.e., to prepare a series of instructions that enables it successfully to complete a computational task) is a skill that many engineers find useful, to the extent that most students now learn to do so early in their engineering curriculum. In addition, engineers should be generally familiar with the broad applicability of computers. This survey of the usefulness of computers to engineers focuses first on the ways engineers use a computer as a tool in arriving at solutions to problems, in the same sense that the hand calculator and transit are tools. Later you will learn how computers are utilized *in* solutions (e.g., as part of a spacecraft guidance system).

The Digital Computer As a Problem Solving Aid

The computer has certainly affected the practice of engineering. It is fast becoming an indispensable design tool, aiding engineers in such ways as these.

Literature search. An engineer is designing a comprehensive computer system for a university. This system is expected to process grades, prepare schedules, handle accounts, maintain student records, do some instructing, and perform a multitude of other useful functions. The engineer surmises that systems to handle various parts of this problem have been developed by individuals at other institutions and companies, but how does he learn who the people are and what they have developed? He could easily spend several weeks tracking down reports of the work done on this sort of thing elsewhere, through library searches, reading, talking, corresponding, telephoning, and traveling. And, after all this effort, he would have missed some worthwhile systems that others have worked out simply because there are so many scattered references and because an exhaustive search would be too expensive and time-consuming.

This is the way it is for most engineering problems. The literature search tasks are staggering. Ordinarily the engineer conducts a reasonable search and then proceeds to solve the problem at hand, accepting the risk that he is duplicating the work of someone who has already solved part or all of the same problem. Under the circumstances, it is uneconomical form him to do otherwise.

But the computer is coming to the rescue, enabling the engineer to conduct his search as follows. He lists key words, the kind of words he would keep in mind if he were conducting a search in the library. In this case he would have two lists.

Keyword List 1	**Keyword List 2**
College	Admissions
School	Billing
University	File keeping
Student	Inventory
	Record keeping
	Registration
	Scheduling

He submits these to his company's computer center to be used for search of its files on computer applications. These files consist of magnetic tapes containing abstracts that describe every computer application made public. The computer goes through these tapes and, for every abstract that contains at least one word from list 1 *and* at least one word from list 2, it prints the abstract and the information the engineer needs to get the details. Within the half hour he can obtain a more comprehensive collection of references than he could possibly afford by other methods. (Here the engineer is employing a computer as an aid in design. In this project a computer will also be part of the system he is designing.)

Such search systems are not in widespread use, but you can see why they are under development. They minimize the costly duplication of research and problem-solving efforts and eliminate the need for arduously sifting through "mountains" of books and technical reports. They do the job quickly and exhaustively.

THINK IT THROUGH

A problem that concerns a number of people is the sizeable gap between the frontier of knowledge in a particular field and what the practitioner typically knows and uses. Mankind is not particularly effective in putting what it learns into widespread use, and this is true of education, medicine, manufacturing, and other fields. Digital computers offer considerable potential in reducing that gap. Can you visualize how?

____

____

____

____

Computers can be of assistance to man in making more frequent use of information that is already known through the use of storage-retrieval systems similar to the one just described. Problem solvers will use more of what is already known when it becomes more accessible.

Data reduction. Engineers often have large volumes of data to be reduced to useful form, for example, hundreds of measurements from an experiment. Calculating averages and measures of variability, curve fitting, statistical tests, and the like, are usually time-consuming and tedious if done by hand. The computer is a natural for such chores.

CHECK IT OUT

What operations on your job either do, or could, benefit from the help of a computer in the area of data reduction?

____

____

____

Responses will depend on your job experience.

Mathematical operations. Most common mathematical operations can be executed by the computer. Some of them, like solving simultaneous equations of many unknowns, take hours or days to do by hand. Unlike the equations ordinarily found in mathematics books, those encountered in practice are often messy to solve. The $y = ax^b$ of the textbook is likely to be $y = 1.878x^{2.3}$ in practice. As a result, computation can become very time-consuming. Try finding the value of x that balances the equation $x = 1.31e^{0.27x}$ to within 1 percent. This will take some time by pencil-and-paper methods. You can see why the computer is so important to engineers; without it such chores are a significant time drain.

Simulation. A fast-spreading technique is digital simulation by computer. Since simulation *is* experimentation on a model, and since the whole operation can be computerized, the engineer can, in this sense, conduct experiments on a computer. (You know why engineers turn to the computer to carry out digital simulations if you have ever done one by hand!)

Iterative optimization. You may well recall that iterative optimization can be time-consuming if done by hand. In fact, optimization by formal iterative methods was not often attempted, even considered, before computers became available. Now engineers can take full advantage of this powerful technique.

The "computer draftsman." Auxiliary equipment now enables an engineer to communicate graphically with a computer. This remarkable and significant development is just beginning to affect the practice of engineering but, within a few years, the impact of *computer graphics* will be dramatic. You can see from Figures 11-15 that this capability has enormous potential.

Figure 11. This CRT enables a computer to display letters and numbers. The user communicates to a computer using a typewriter keyboard or the light pen he holds. This physician can indicate which of the tests listed on the screen he wants performed on a patient by pointing with the pen.

Figure 12. This combination of computer, CRT, man, and light pen can accomplish remarkable things. Man can "draw" on the face of the tube with the pen, and the computer can remember what is drawn and reproduce it on the screen whenever it is instructed to do so. Even better, the computer can manipulate these drawings. For example, it can show the object from different perspectives, rotate it, or enlarge it, as instructed by the pen and the keyboard. In this particular view, the operator has pressed the "line deletion" button, which enables him to point to any line and have it erased from the screen and the computer's memory.

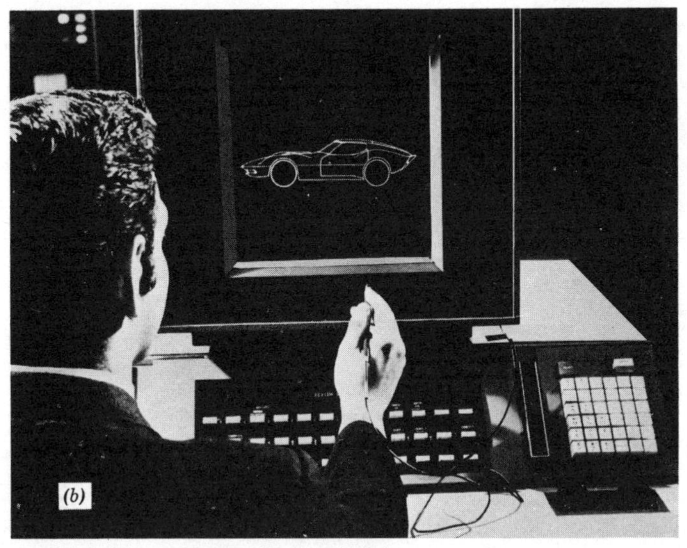

Figure 13. These before and after photographs of the screen pictured in Figure 12 illustrate some of this system's capabilities. When the engineer finishes his work and wants to make a permanent record of what appears on the screen, he does so by instructing the system to store it, in which case it can be called back to the screen at any time. He can also have the computer instruct a plotting machine, which will prepare an inked drawing, or he can obtain a permanent copy by photographic means.

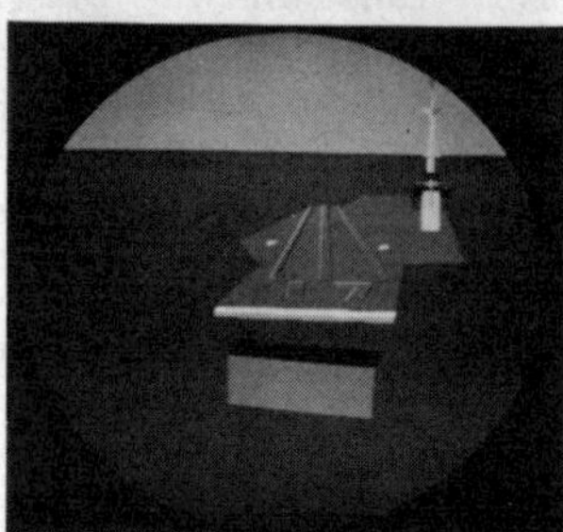

Figure 14. Given a series of numbers that locates the corners of an object as X-Y-Z coordinates, this computer system will prepare television pictures of that object at any perspective and to any scale. Since it can change view up to 30 times a second, it can give the illusion of movement. Views like those shown here can be made to change "continuously" in response to movement of typical aircraft controls, thus allowing a pilot to practice landings on this carrier–in color, at that. Obviously this system is useful for simulation.

Figure 15. The computer system referred to in Figure 14 can do "visualization wonders" for engineers and their clients. Given a minimum of information in terms of coordinates, the computer can "take you for a ride" along a stretch of highway that, at the moment, exists only as a concept. This photo was taken directly from a CRT screen during such a ride. The same can be done for a building–it can be viewed from any prospective as it would appear to an external observer, and views of the external world can be displayed as they would appear to an observer within the building. These feats give you some idea of the exciting things that are coming in the field of computer graphics.

In this sampling of the ways that engineers use computers, there are applications that are close to commonplace and others that are just coming over the horizon. While gaining insight into the usefulness of computers in engineering, you are getting a good idea of what they offer to architects, businessmen, physicians, and other problem solvers.

CHECK IT OUT

How are computers used to solve problems on your job? What ways could they be used?

____

____

____

____

Responses depend, of course, on your job experience.

Utilization of Computers *in* Solutions

In addition to aiding the designer as he arrives at a solution, a computer often becomes a part *of* his solution, for one or more of the following reasons.

1. The solution requires a means of *storing and retrieving* information, and a computer is the most effective alternative.

2. The solution requires a means of *processing* information, and a computer is the most effective alternative.

3. The solution requires a means of handling information quickly, and *only* a computer is fast enough.

4. The solution requires a means of keeping track of many concurrently changing, interacting events or variables, and *only* a computer is capable of it.

Elaboration follows, but first, note the generality and usefulness of this classification of computer applications; it is the pattern of applications in education, business, government—everywhere.

1. Information storage and retrieval. The preservation of knowledge so that it can be retrieved without unreasonable time and expense is a problem in many fields of endeavor. The story is the same in medicine, law, business, education, and government, as well as in engineering. Take medicine; with the thousands of periodicals, books, and reports that fill libraries, how does a physician learn quickly whether anyone anywhere has successfully treated a case of Ozarkmumblitus? Problem solvers in other fields have related information search problems; brains, books, and file drawers—familiar means of storing information—no longer suffice. But the computer has a memory that is remarkably reliable, large, and fast. It offers real hope.

THINK IT THROUGH

Can you think of any problem that state legislators, judges, and administrators would have in common and which could be alleviated by the use of computers?

____

____

An answer to this exercise appears in the following paragraphs.

Here is a sample application. Representatives of a state government requested a consulting firm to develop a more effective system for storing and searching the state's legal information, which includes all statutes and all court decisions. The heart of the solution to this problem is a computer. With this system, lawyer, judges, and legislators can search the state's statutes in order to isolate all laws pertaining to a given subject. If a legislator wishes to know what laws have something to say about the education of handicapped children, he prepares keyword lists and submits them to the computer on punched cards. The computer subsequently prints the numbers of all statutes that contain the words *education, handicapped*, and *children* (or synonyms of these words). In fact, it will print the relevant statutes themselves if so instructed.

This class of applications, illustrated by these medical and legal storage-retrieval systems, is referred to as the computer's library function, since it concerns general knowledge, much of which is available in libraries. But man has a second major type of information storage-retrieval problem. It involves private information (e.g., an insurance company's files on its policyholders) now typically stored in the office files of corporations, government agencies, schools, hospitals, and numerous other institutions. Computers are well suited for this type of storage-retrieval task also. It is called the filekeeping function.

TECHNICAL TERMS

Write a brief definition for each term.

1. Library function (of a computer) ______________________________

____

2. Filekeeping function (of a computer) __________________________

____

1. The library function of a computer refers to a class of applications in which the computer stores and retrieves information much in the same way that a library and library research might operate, but at much faster and more efficient rates. This term generally refers to the storage and retrieval of public information such as laws and public documents or other published material, the kind of information one would find in a public library.

2. The filekeeping function of a computer refers to a class of applications in which the computer stores and retrieves information much in the same way that a filing system operates in a company or private organization, but at much faster and more efficient rates. This term generally refers to the storage and retrieval of private information, the kind of information one would find in the files of a company or organization.

A hospital has a staggering filekeeping problem. Much time is spent in recording, filing, exchanging, and looking up information. Then there are the thousands of square feet consumed by patient records. A team of engineers called on to solve a hospital's "information problem" proposed the computer system summarized by Figure 16. The benefits are impressive. This system keeps an up-to-the-minute record for each patient and answers inquiries within a few seconds. A sampling of the services it provides: checks prescriptions for inconsistencies and errors, reminds nurses when it is time to administer medication; and tells the kitchen how many pounds of each type of food on the menu will be needed for the next meal.

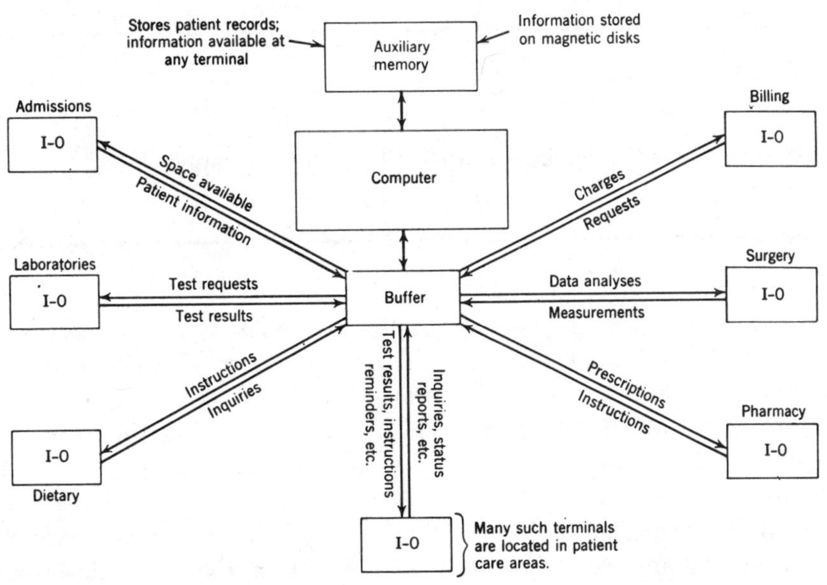

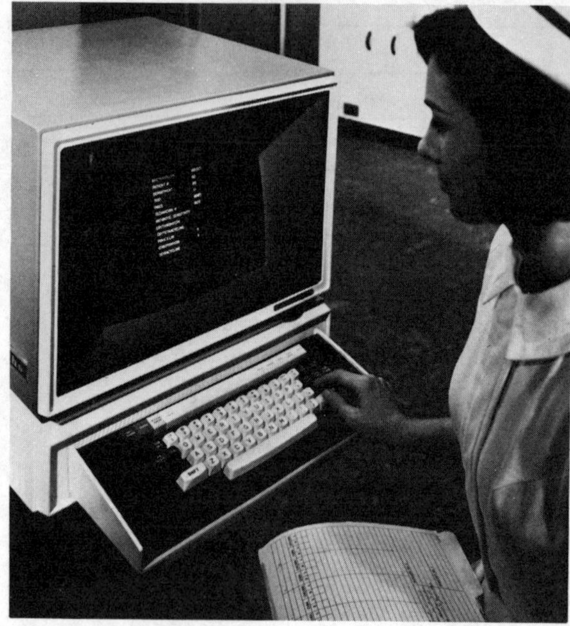

Figure 16. A "computerized" master record-keeping system for a hospital. From any of the terminals located throughout the hospital, an authorized person can add information to a patient's file or request information from it. Examples of data this system stores, processes, and transfers are indicated by the arrows. Equipment at a typical input-output station is pictured by the inset. This computer can also serve as the heart of a patient-monitoring system that alerts hospital personnel when there is a change in the vital life signs of a critically ill person.

THINK IT THROUGH

Why wouldn't such a computer system, with some thirty terminals hooked up to the computer, result in intolerable delays? (*Hint*: If you are not familiar with computer operations, think of the simple example of typing out a tape on the teletype before sending out the message in order to make sure that the transmission takes as little time as possible.)

____

____

____

The answer to this exercise appears in the following paragraphs.

Delays of more than a second or two are very unlikely in view of the computer's speed and of the manner in which the system operates. It scans all terminals many times a second in round-robin fashion. When it encounters a terminal at which there is a customer (call him user A), the computer devotes 1/20 second to his request (which may be to file, look up, or process information). If it finishes user A's job in that interval, as is likely, fine. If it doesn't, it puts his job on ice, so to speak, and it scans to see whether there are other active terminals. Scanning takes negligible time. If there are no other terminals at which someone has a task, the machine comes back to user A and gives him another 1/20 second. Then it scans again. But suppose that this time it finds another active terminal, at which user B has work for it. It gives user B 1/20 second, scans and, finding no new active terminals, it returns to user A. And so it goes. Even if most of the 30 terminals are active simultaneously, which is very unlikely, most users' tasks will be completed quickly enough to give each the impression that he has exclusive use of the computer.

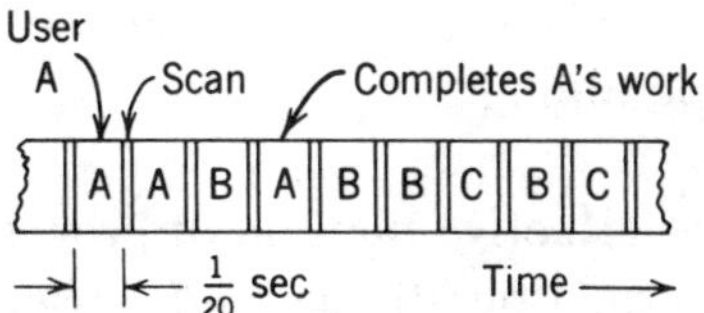

This mode of operation, called *time sharing*, is a relatively recent development in the computer field. You will hear a lot about it and surely feel its impact. To believe that a system like this exists requires that you have a feel for the quickness of the computer. Actually, it is serving its customers sequentially, but it switches so rapidly and completes its work so quickly that it appears to be serving them simultaneously.

The communication buffer between man and computer (Figure 16) is an ingenious device without which time sharing would be impractical. User *A* spent 7 seconds typing his request for patient information. Fortunately, during that period, his typewriter was not transmitting directly to the computer. His message went to the buffer, where it was stored until user *A* pressed the end-of-message key. Then, when the computer scanned to user *A*, the buffer fired his message into the computer in a few milliseconds. This frees the very fast and expensive computer from the relatively slow process of typing. The buffer, then, serves as a "time compressor" for incoming messages. It serves also as a "time expander" in the reverse direction, so that the 0.2-millisecond message it receives from the computer is fed to the typewriter at a speed the latter can follow.

CHECK IT OUT

Are computers applied in library or filekeeping functions on your job? Could the introduction of such applications, or expansions of present ones, be beneficial for your organization? Give examples. Explain.

Responses depend, of course, on your personal experience, but some hints at the multiple possibilities of computer applications in the library or filekeeping function appear in the following paragraph.

This description of a computerized master information system for a hospital is fiction; some parts of it are operating at some hospitals, but nothing like this comprehensive "manager of medical information" currently exists. However, you have had a glimpse of what is coming not only for hospitals but for insurance companies, motor vehicle bureaus, local governments, mail-order houses, school systems, and so many other enterprises in which the record-keeping problems are compounding.

2. When the computer is the most effective method of processing the information at hand. In the preceding applications, information storage and retrieval were primary, and processing of information was incidental. Here the reverse is true. Probably most computer applications you know of are of this type, which is understandable; this *is* the most frequent type of application.

A computer is likely to be the cheapest way of doing the job whenever large numbers of repetitive operations must be performed. This is why a computer is called on to prepare an electric utility's 120,000 bills each month, to schedule a university's 38,000 students, to prepare a corporation's 17,000-employee payroll, and to process the Census Bureau's statistics on millions of citizens. Understandable.

LET'S BE SURE ABOUT THIS

A computer certainly is not always the most effective method of processing information. What is the key to making the processing of information an efficient way to use a computer (and to making the use of a computer an efficient way to process information)?

____

____

____

The key here is repetitiveness: volumes of data, millions of calculations; cyclic execution of the same basic operations on each piece of data.

3. When only the computer is fast enough. In the preceding types of computer applications, man and machine competed for the task at hand, and the computer got the job because it excelled at the task in question, on economic and perhaps other grounds. But there is no choice here; man cannot respond fast enough. Here is such an instance.

What could be an ICBM is detected by radar, apparently headed toward the United States. There is no time to spare. The nature of this object must be verified and a decision made concerning interception in a matter of seconds. Unfortunately, there is no time to fly out and have a look at the object for identification purposes. Those days are long gone. The only practical way now is to track the object by radar and on the basis of this information compute its location, speed, and probable destination, and then quickly see whether it can be accounted for on the basis of aircraft and spacecraft scheduled to be in the area. Based on this information, a choice must be made: forget it, wait, intercept. All of this must be accomplished in a few seconds; only a computer is capable of such speed. If an interceptor is to be launched, aiming it constitutes another challenging problem that only the computer can handle (Figure 17).

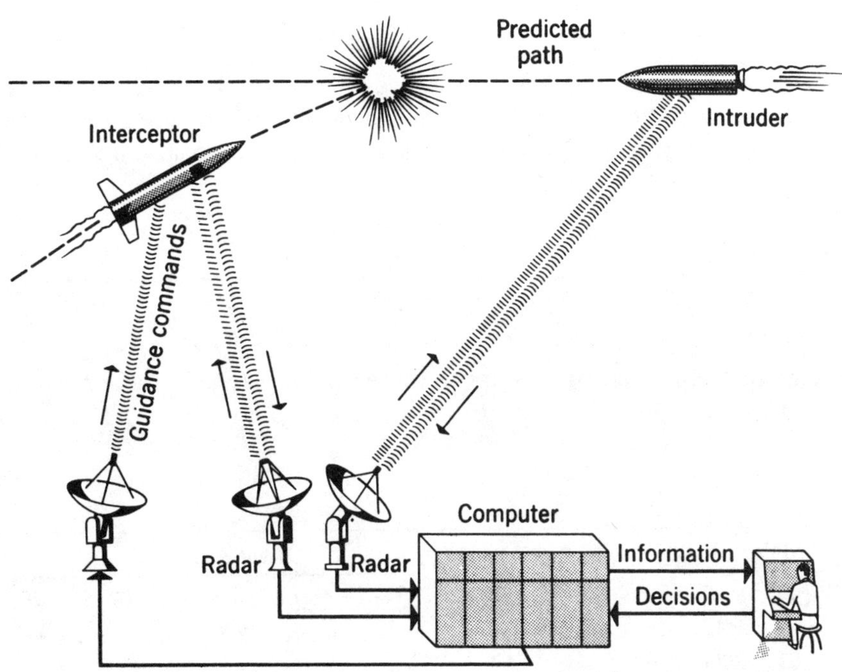

Figure 17. To intercept and kill an intruding ICBM, the path of the intruder must be predicted and an intercept point calculated, which is likely to be hundreds of miles from its present position. Hence the interceptor has a computed path that it must follow to the point of kill. Unfortunately, there are forces that tend to steer it off course. Therefore it is necessary to sense deviations from the intended path, compute necessary corrections in course, and transmit these guidance instructions to the interceptor. This sequence of events–sense position, compute corrections, transmit–goes on continuously. This guidance system must react in microseconds in view of the missile's 29,000 km/hr speed. While all this is going on, the radar continues to sense the intruder's path and target. Therefore, the interceptor must be guided to a point of kill that, itself, is changing. And that's not all. An ICBM would no doubt be accompanied by clouds of decoys. The real things must be sorted out in a hurry. Such matters complicate the whole process considerably. One thing is for sure: buildings full of men could not possibly make these computations in time.

LET'S BE SURE ABOUT THIS

In cases such as the interception of an ICBM, what is the most conspicuous need that dictates the use of a computer?

Conspicuous in this type of computer application is the speed with which things happen. Certain manufacturing processes like the rolling of steel slabs into flat sheets operate at very high speeds, making it impractical for humans to control them. In an increasing number of such instances, computers are called on to do the job.

4. Too many things to keep track of. Suppose there are 30 planes in the air around an airport. How can a man keep track of them? He can't! But a computer can. (See Figure 18.)

	Flight A715					Flight A715					
	$T-2$	$T-1$	T	$T+10$	$T+20$	$T-2$	$T-1$	T	$T+10$	$T+20$	$T-2$
x	9.12	9.03	8.94	8.49	8.04	9.03	8.94	8.85	8.40	7.95	8.94
y	27.04	27.03	27.02	26.97	26.92	27.03	27.02	27.01	26.96	26.91	27.02
z	2.14	2.14	2.14	2.14	2.14	2.14	2.14	2.13	2.08	2.03	2.14
	Positions 1 and 2 seconds ago		Position now	Predicted positions							
	$T = 9{:}22{:}02$					$T = 9{:}22{:}03$					

Figure 18. Once a second, the computer receivers information from radar on the position of each plane in its traffic control region. It stores this information in the form of X-Y-Z coordinates. What you see here is the computer's "file" on Flight A715 for two successive seconds. The computer is doing likewise for other planes under its control at the time, switching its attention rapidly from one plane to the next. The computer can do more than simply store positions. For example, it can predict where each plane will be in the near future, as shown in the predicted position columns. It does this by using the present and immediate-past positions of the plane to compute its speed and heading. Then on these bases it computes where the plane will be 10 seconds and 20 seconds later. In this way the computer can foresee potentially hazardous situations and alert traffic control personnel. Traffic controllers can request the computer to display this information graphically on their scopes, including the predicted positions, for specific planes or all planes in their area.

THINK IT THROUGH

Does the computer keep track of all planes simultaneously and continuously? Explain. (*Hint*: Think back to the hospital time-sharing case outlined above.)

An answer to this exercise appears in the following paragraph.

For all practical purposes, the computer keeps track of all planes continuously and simultaneously. Actually, however, it concentrates on one plane at a time, switching its attention from plane to plane and updating its knowledge of the whereabouts of each one at more than satisfactory frequency. This is the high-speed scan process that is characteristic of time-sharing computer systems.

CHECK IT OUT

Are computers applied on your job or in your organization in cases where the computer is the most effective way to process information at hand? In cases where only the computer is fast enough to do the job required? In cases where there are too many things to keep track of effectively without a computer? Could introduction of such applications, or expansion of present ones, be beneficial for your organization? Give examples. Explain.

Responses depend, of course, on your personal experience, but some thoughts along these lines appear in the following paragraphs.

Some manufacturing processes parallel the air traffic control situation in that they involve dozens of variables (temperatures, pressures, speeds, etc.) that are subject to frequent change and are interrelated. No human, or any number of them, can cope with all of these at once. Can you imagine what is involved in trying to keep track of 125,000 freight cars in a railroad system? Conspicuous in these situations is a large number of changing, interrelated events or conditions—too much for a human mind to grasp at one time. For the computer, this is no problem.

And so, in some instances, there is no choice; only a computer can handle the job. But in most applications, it is a matter of effectiveness—alternative methods have been compared, and the computer was found to be superior. The designer of the utility company's billing system compared the total costs of preparing bills by desk calculators, accounting machines, and computer and found the latter to be the cheapest.

If you have become familiar with the concepts in this learning segment, you have a good overview of the manner in which engineers in particular and man in general are utilizing digital computers. With this overview in mind, you are better equipped to spot a potentially profitable computer application.

SELF QUIZ

1. What two inputs are essential for a computer operation, regardless of the device used to accomplish the input?

---

2. What two functions must be present in the central processing unit of a computer for it to complete its processing activities?

---

3. List the essential steps that must be followed in order for a computer to complete a job successfully.

---

4. Define the term *data*.

---

5. Define the term *program*.

---

6. Define the term *memory*. Be sure to make reference to two kinds of computer-system memory.

---

7. Define the term *branching*.

---

8. List six ways that engineers use computers in the process of arriving at solutions.

9. List the four basic ways in which computers are needed as a part of the solutions to problems.

10. How are library functions and filekeeping functions of computers the same? How are they different?

1. Essential inputs for a computer operation are data, or information, and a program that tells the computer what to do with data.

2. The central processing unit of a computer must have a functioning arithmetic processing capability and a functioning working memory. (If you have some knowledge of computers, you would probably add control function.)

3. Five essential steps in the operation of a computer are

1) Prepare or select a program.

2) Enter the program into the computer's working memory.

3) Communicate the data to the computer's working memory.

4) The arithmetic unit processes the data according to instructions contained in the program.

5) The results of the processing, the output, are communicated to the user of the computer operation.

4. Data are all the items of information fed into a computer, processed by the computer, and communicated, in processed form, to the user of the computer.

5. A program is a set of instructions that is fed into a computer and stored in its working memory. It tells the computer what to do with data fed into the computer.

6. The term memory refers to the computer's ability to retain information. There is a memory internal to the central processing unit of a computer that is called working memory. There is also memory external to the central processing unit that is called auxiliary memory; this is in the form of magnetic tapes or disks that can be put in communication with the central processing unit.

7. Branching is a computer's ability to compare numbers or letters and follow a different series of instructions, depending on the results of the comparison.

8. Engineers use the computer as a part of the problem-solving process in the following ways.

To search the literature

For data reduction

To carry out mathematical operations

For simulation

To carry out iterative optimization procedures

To utilize the aid of computer graphics

9. Four basic needs that may require the use of a computer in a solution are

The need to store and retrieve information efficiently.

The need to process information efficiently.

The need to process information more quickly than can be done by any other means.

The need to keep track of more information and act on it faster than is possible with any other means.

10. The library and filekeeping functions of a computer are both storage and retrieval functions. The library function is the storage and retrieval of public information such as might be found in a public library. The filekeeping function is the storage and retrieval of private information such as might be found in the files of a company or other private organization or individual.

Learning Segment 12
SOME GENERALIZATIONS

LEARNING OBJECTIVES

- Describe the differences between pure science, applied science, and engineering.
- Become aware of some of the specialties in engineering.
- Know some of the nontechnical areas of knowledge that are important for the practice of engineering activities.
- Become aware of some of the historical developments that have contributed to contemporary engineering.

INTRODUCTION

Before generalizing, here is one more brief example of the types of things engineers must know. In this case, it is what is required to design a transformer. A transformer generally consists of two current-carrying coils electrically insulated from each other and wound on the same iron core (Figure 19). An alternating current in the primary coil produces an alternating field of magnetic flux that passes through the metal core to the secondary coil in which an alternating current is induced.

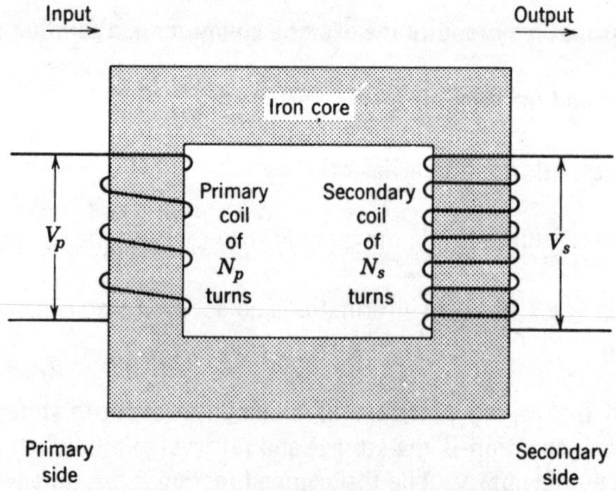

Figure 19. A simple step-up transformer. It transforms the primary voltage (V_p) to a higher secondary voltage (V_s).

In a basic physics course, an engineering student usually learns something about transformers, under the subject of electromagnetism. However, the coverage is ordinarily limited to the principles of ideal transformers. Probably the only quantitative relationship he learns is the one between the primary and secondary voltages, described by this simple model. But that's about all. If he is assigned problems to work he is given numerical values for three of these variables and asked to compute the fourth (e.g., given values for N_p, N_s, and V_p, find V_s). But the engineer must know considerably more than this to *design* transformers.

$$\frac{V_p}{V_s} = \frac{N_p}{N_s} \qquad (a)$$

where V_p and V_s are the primary and secondary voltages.

N_p and N_s are the number of windings in the primary and secondary coils.

THINK IT THROUGH

Can you think of any knowledge the engineer must have, beyond a knowledge of the ideal transformer, in order to design transformers?

____

____

____

____

____

____

Answers to exercise appear in the following paragraphs.

For one thing, the engineer must extend his knowledge beyond the ideal case. Therefore, in a subsequent applied science course, probably in his sophomore year, he learns more about the types of losses that were assumed away in the study of the ideal transformer. There are realities such as resistance of the wire in the coils, flux leakage, eddy currents, and hysteresis, which collectively cause the output of a real transformer to be less than the ideal. Again, however, if he is asked to solve transformer problems in such a course, he will most likely be given numerical values for all but one of the variables, including losses, and be asked to solve for the one unknown. But certainly it is not that simple in practice.

THINK IT THROUGH

In a college course in applied science, the engineering student will be given all the information about all of a problem except the value of one variable. But that is not the way things happen in real life on the job. What information is the engineer likely to be given when an assignment to design a transformer comes across his or her desk?

Answers to this exercise appear in the following paragraphs.

On the job he will usually be told that a transformer is needed, with an output specified in terms of the current (I_s) and voltage (V_s). He may also be told the primary voltage (V_p) and the alternating current frequency (f), but not much else. He takes it from there. But many of the variables like N_p, N_s, and I_p are as yet unknown; so are the size and shape of the core, the material from which it is to be made, and a number of other unspecified solution variables. *This* is a lot different than the relatively simple "plug in all the values but one and crank out the answer" routine that sufficed in introductory courses. Here the engineer has many interdependent unknowns; how can he assign values in such a situation?

THINK IT THROUGH

Do you have any ideas on how the engineer can assign values to the many unknowns in the transformer problem presented?

Answers for this exercise appear in the following paragraphs.

The details of his procedure are not important here but, in general, it goes like this. The designer has an empirical equation, available in an electrical engineering handbook, that enables him to estimate the number of windings needed in the secondary coil, *if* he is willing to make some assumptions. So, for a starter, he must assume trial values for A (the cross-sectional area of the transformer's iron core) and B (the density of the magnetic flux flowing through the iron core). There are rules of thumb in practice to guide his selection of trial values for A and B. Although these are only "ball-park" values, they do enable the engineer to get a foothold in the cyclical process of successive approximations.

$$N_s = \frac{(V_s)\,10^8}{4.44\,BAf} \qquad \text{(b)}$$

Now, using the values given for V_s and f and the values assumed for A and B, he can use equation (b) to compute a first approximation of N_s. Now he has values for V_s, V_p, and N_s, and using these in equation (a), he can calculate N_p.

After making tentative selections of materials and the coil-core configuration, he can estimate the resulting losses. But note that many of these variables are interdependent; B, for which he originally assumed a value, depends on many of the subsequently established characteristics. So about now he starts the process over again, but this time with a better basis for setting values of A and B. Thus he goes through a series of successive approximations until all features appear satisfactorily specified.

Even though you may not follow all of this, if should be apparent that there is much more involved than the simple plug-in process associated with the use of equation (a) in introductory physics. There are few equations and many interrrelated unknowns, so that design of a transformer turns out to be a cyclic, cut-and-try process, in which the engineer relies heavily on empirical knowledge in the form of formulas, rules of thumb, tables, and graphs. Some of this empirical information is acquired from upper-class engineering courses, but most of it comes from on-the-job experience and handbooks.

YOU DECIDE

What is your reaction to the following definition of engineering?

Engineering is the application of science.

Nonsense! Of course, engineers use science in the solution of problems, but much of what they rely on is empirical (recall the transformer case) or is under the general heading of engineering systems. Furthermore, those who consider engineering to be applied science are ignoring the fact that engineering is still very much an art.

SOME GENERALIZATIONS

Basic Physical Science

A very important part of an engineering education concerns the physical sciences, primarily physics and chemistry, as evidenced by the number of courses in these subjects in the typical engineering curriculum. In order to develop complex devices, structures, and processes, one must acquire a *fundamental* understanding of the laws of motion, the structure of matter, electromagnetism, the behavior of fluids, the conversion of energy, and many other phenomena of the physical world.

But knowledge of basic physical science is hardly enough. It takes a lot more than familiarity with ideal transformers to design real ones successfully. This is true for all engineering creations. People do not want their bridges designed by engineers who know only basic physics and chemistry any more than you want to be treated by a physician whose only qualification is knowledge of basic physiology. There is a big step between the basic principles of physical science and technical solutions to real problems. An engineer's formal education must equip him or her to bridge this gap. Therefore, after the rigors of basic physical science, engineers must study *applied* physical science, engineering systems, and a great accumulation of empirical know-how.

Applied Physical Science

Applied science is the "how" and "where" of applying basic principles. An example of applied physical science is circuit analysis, which concerns the application of knowledge of fundamental electrical phenomena (charges, electromagnetic waves, electron flow, etc.) to the understanding of basic electrical circuits. It is in such a course that the engineering student extends his or her knowledge of transformers from the ideal to the actual case. Other applied physical sciences are taught under course titles such as thermodynamics, mechanics of solids, fluid mechanics, and properties of materials. These are concentrated in the middle portion of the typical engineering curriculum.

Engineering Systems

Engineers are creators of systems ranging in size from microsystems like the electrical circuits seen only through a microscope to the supersystems exemplified by our telephone network. The transistor, toaster, automobile, computer, airplane, dam, oil refinery, highway network, and all other engineering creations *are* systems. To design such systems, engineers must understand them thoroughly and know how they are synthesized. Thus, in courses in the junior and especially the senior year of an engineering curriculum in college, students have learned to design communication, transportation, energy conversion, control, manufacturing, and other types of systems.

THINK IT THROUGH

It is beginning to sound as if engineers are superhuman. Do you suppose that any one engineer studies all these systems? Why, or why not?

____

____

An answer to this exercise appears in the following paragraph.

No student studies *all* of these. It is customary to specialize to some degree, primarily because large and substantially different bodies of knowledge are required by basically different types of problems. It is virtually impossible for an engineer to be competent in designing bridges *and* television equipment *and* jet engines *and* metal refineries *and* textile machines. As a consequence, some specialization is inevitable. Therefore, during the latter part of an undergraduate program, the student will major in some branch of engineering. The choices are many; some major ones of longstanding importance are:

- *Aerospace* engineering. Primarily the design of systems for travel above the earth's surface. Examples: aircraft, spacecraft, air cushion vehicles, guidance and other systems associated with flight.

- *Chemical* engineering. Primarily the design of processes for the chemical transformation of materials on a large scale. Example: facilities for the production of gasoline, paint, explosives, rubber, cement.

- *Civil* engineering. Primarily the design of major structures *and* the means of constructing them. Examples: highways, bridges, dams, canals, water supply and sewage diposal systems, airports, and waterports.

- *Electrical* engineering. Primarily the design of means by which electrical energy is created, transferred, and used. Examples: electrical generators, transmission networks, communication systems.

- *Industrial* engineering. Primarily the design of operating systems for producing goods and services. Examples: automobile plants, printeries, shipyards, hospitals (not the building itself).

- *Mechanical* engineering. Primarily the design of systems by which energy is converted to useful mechanical forms. Examples: combustion engines *and* the mechanisms required to convert the output of these machines to the desired form, including pumps, compressors, and transmission systems.

LET'S BE SURE ABOUT THIS

Without going back over the preceding list, try to describe the general nature of things designed by each of the following types of engineers *and* give one example of the kind of things each type of engineer designs.

1. An aerospace engineer designs ______________________________

______________________________,

for example, ______________________________.

2. A chemical engineer designs ______________________________

______________________________,

for example, ______________________________.

3. A civil engineer designs ______________________________

______________________________,

for example, ______________________________.

4. An electrical engineer designs ______________________________

______________________________,

for example, ______________________________.

5. An industrial engineer designs ______________________________

______________________________,

for example, ______________________________.

6. A mechanical engineer designs ______________________________

______________________________,

for example, ______________________________

______________________________.

1. An aerospace engineer designs systems that travel above the earth's surface, for example, aircraft, spacecraft, air cushion vehciles, guidance and other systems associated with flight.

2. A chemical engineer designs processes for the chemical transformation of materials on a large scale, for example, facilities for the production of gasoline, paint, explosives, rubber, and cement.

3. A civil engineer designs major structures and the means of constructing them, for example, highways, bridges, dams, canals, water supply and sewage disposal systems, airports, and water ports.

4. An electrical engineer designs the means by which electrical energy is created, transferred, and used, for example, electrical generators, transmission networks, and communication systems.

5. An industrial engineer designs systems for producing goods and services, for example, automobile plants, printing plants, shipyards, health delivery systems in hospitals.

6. A mechanical engineer designs systems by which energy is converted into useful mechanical forms, for example, combustion engines and the mechanisms to convert the output into desired form, including pumps, compressors, and transmission systems.

It is primarily in the third- and fourth-year design courses that education in the branches of engineering differs. The student of electrical engineering studies the behavior and design of electrical machines, communication systems, power distribution networks, and the like, while the student of civil engineering studies structures, water supply systems, city planning, and related subjects. One type of system that they all study, regardless of specialty, is the feedback control system, because most engineering creations include such systems.

Although specialization along traditional lines is still common in engineering *education*, most problems encountered in practice require knowledge from two or more of the traditional engineering specialties. Design of a commercial chemical process such as a plant for converting oil-sands into crude oil requires knowledge that is traditionally a part of the education of a chemical engineer, as well as knowledge acquired by civil, electrical, industrial, mechanical and metallurgical engineers. As a result, an engineer must often work closely with engineers educated in specialties other than his own, and must himself employ some of the knowledge from other branches of engineering. Thus, the engineer typically finds that, on the job, his knowledge must extend across the traditional specialty boundaries.

Codified Empirical Knowledge

This is an accumulation of ideas, standard practices, formulas, rules of thumb, and the like, some of which is published and some of which is retained in heads, files, and notebooks. The nature and role of this practical knowledge is apparent in the transformer case. Such information is not based on scientific principles. It is cumulative know-how that has evolved over a long period, during which it has proven sound and generally useful. Engineers rely heavily on this "communicated experience," and so this type of knowledge is woven into upper-class engineering studies.

Other Knowledge

There are some important nontechnical aspects of your intellectual development; certainly knowledge of physical science and engineering will not suffice.

THINK IT THROUGH

What are some of the nontechnical areas of knowledge that an engineer requires?

___

___

Answers to this exercise appear in the following paragraphs.

The practice of engineering requires some familiarity with economics, political science, psychology, and sociology. This breadth of knowledge is important for a number of reasons; for instance:

- Engineers must know the "economic facts of life." To be of value to employers and to benefit society, they must be aware of the importance and intricacies of costs, price-demand relationships, return on the investment, depreciation, and other economic realities. They will be heavily involved in economic decisions.

- Engineers will be working with persons in many fields of endeavor: economists, accountants, politicians, sociologists, psychologists, lawyers, and union leaders. They should be aware of the contributions that these people can make, be able to talk with them intelligently, work with them, and understand their problems.

- Educational breadth equips and motivates engineers to show greater concern for the society they affect. What better reason is there for insisting that education not be exclusively technical?

For reasons such as these, a minimum of one-fifth of an undergraduate engineering education is reserved for study of the humanities and the social sciences.

In summary, an engineer's creations are based on an accumulation of knowledge varying from pure science to pure empiricism, acquired in science, applied science, and engineering-systems courses, as well as through experience. In general, the typical solution owes less to science and more to other types of knowledge than many would lead you to believe. It is worth making an issue of this, partly to refute the common impression that engineering is (or is primarily) the application of science.

SCIENCE AND ENGINEERING: THE DISTINCTION

True, science and engineering are closely related and, in some respects, are interdependent. On the other hand, there are some fundamental differences between them that elude many people partly, no doubt, because of this very relationship. Science and engineering differ in four major respects: *the basic processes characteristic of each; predominant day-to-day concerns; primary end-product;* and *knowledge employed.*

Science is a body of knowledge, specifically, man's accumulated understanding of nature, *and* the means by which that understanding is improved. *Scientists* direct their efforts primarily to extending this knowledge. In their search for better understanding, scientists engage in a process called *research.* In so doing they devote much of their time to hypothesizing and refining models (explanations, classification schemes, mathematical representations, etc.) of natural phenomena and to evaluating those models on the basis of observations from experiments and the world around them. There is certainly more to the work of scientists, but the intent here is mainly to highlight their heavy involvement with models: model development, model refinement, and model verification. Now contrast this with the engineer's design process in which modeling is strictly a means to quite different ends.

THINK IT THROUGH

What are an engineer's ends? Contrast them with a scientist's.

____

____

____

____

Answers to this exercise appear in the following paragraph.

These ends, in contrast to knowledge, are tangible, generally in the form of physical systems created to satisfy some human need or want. Let there be no mistake—the weather satellite, the radio telescope, the heart-lung machine, the nuclear power station, the electronic computer, and the artificial kidney evolved from engineering endeavors. Engineers develop these contrivances through a creative process referred to as *design* (in contrast to scientists' central activity—research). Some of our prime concerns as designers are economic feasibility, safety, public acceptance, and manufacturability of solutions. These are hardly major concerns of scientists. Instead, scientists are more concerned about validity of their theories, reproducibility of their experiments, and adequacy of their methods of observing natural, phenomena.

But it's not only objective, methodology, and day-to-day concerns that distinguish these two disciplines; even their knowledge needs differ. This can be illustrated by contrasting what scientists and engineers must know about control systems.

THINK IT THROUGH

Can you describe the primary difference between the scientist's knowledge of the principles of control systems and an engineer's knowledge of the such principles?

____

____

____

____

An answer to this exercise appears in the following paragraphs.

Scientists, particularly in the biological and social sciences, should have some knowledge of control systems, but certainly not because they design them. Their primary interest is in *explaining* existing control mechanisms in nature (e.g., the means for controlling blood supply in the body). In contrast, the prime objective of engineers is to *create* systems that fulfill control needs in man-made systems (e.g., *given* that steam pressure is to be maintained at a special pressure, to design a system that does so).

Similarly, a comparison of what a student learns about how transformers operate in a physics course and what an engineer must know to design them can be instructive. Note that in a physics course a student learns how to determine what the result will be, given various voltage, current, coil, and other conditions. In contrast, the engineer must determine characteristics of the coil, input, and the like, *given the desired result* (*output*). Thus, in engineering the emphasis is on *creation*, whereas in science the emphasis is on *explanation* (some like to express this crucial difference in terms of a synthesis and analysis).

Faraday's formulation of the principles of electromagnetic induction was a contribution to science. The use of this knowledge in the design of electric generators is engineering. When man came to understand nuclear fission in the late 1930s, this was an important scientific discovery. Applying this knowledge in the design of power-generating nuclear reactors is engineering. This is not to say that engineers do not do research in the course of finding solutions to their problems. Engineers who developed practical means of converting brackish water to usable form performed some research to gain additional knowledge about the fundamental processes involved. However, their goal was development of an economical conversion process; they engaged in research only in order to solve that problem. When a space vehicle reenters the earth's atmosphere at very high speeds, the heat generated will melt any known metal. It was necessary therefore for engineers designing reentry vehicles to engage in research to find a material capable of withstanding this intense heat. The knowledge that resulted is a by-product of their efforts to develop a successful reentry vehicle.

So engineers do research, and certainly scientists design instruments and solve problems. The key to the distinction is what constitutes a prime objective and what constitutes a means to an end.

Figure 20. This is by far the largest radio telescope. It is a wire mesh reflector supported in a natural hollow in the mountains of Puerto Rico. Radar signals originate from the movable transmitter and are reflected toward outer space. These signals bounce off planets and stars and return as echoes that are focused by the reflector on the suspended receiver. These echoes are then analyzed by scientists to yield new knowledge about the universe.

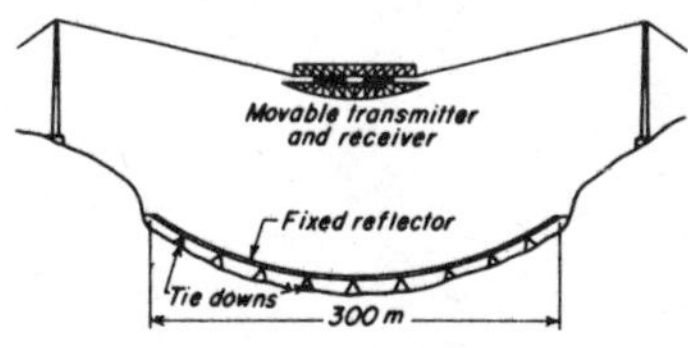

YOU DECIDE

Is the radio telescope in Figure 20 a scientific or engineering achievement?

____

____

This radio telescope is a prime example of an engineering creation that the press typically calls a scientific achievement. To be precise, it is a scientific instrument used in radio astronomy, the design and construction of which are engineering achievements. Engineers determined the site for this telescope; they designed it; in fact, an engineer conceived the basic idea. I mention these things not because scientists and engineers quibble about credit due, but because the types of work involved in design and use of this instrument are quite different.

Turning From Widely Held Misconceptions

While there are some myths surrounding engineering, there are also the inevitable little-known truths. The next unit is addressed to a very important one; the role of the engineers as an agent of social change, a fact that seems to elude many engineers as well as the public. Yet, through their creations, they precipitate more social change than those who work at it, although they do so indirectly, unintentionally, and sometimes unwittingly. This role as a social revolutionary—precipitator of social transformation—is worth examining in detail.

SELF QUIZ

1. Why is it *not* accurate to call engineering the application of science?

2. Explain how a problem presented in a basic physical science course would differ from a problem in the same subject area presented in an applied science course. Explain how the same problem area would differ as presented to an engineer in practice.

3. Give a brief description of the kind of thing designed by each of the following types of engineers.

Aerospace engineer

Civil engineer

Electrical engineer

Industrial engineer

Mechanical engineer

4. List at least three nontechnical areas of knowledge that are important for an engineer.

5. List four ways in which science and engineering differ.

1. Science and engineering differ in a number of ways, four of which are listed in the answer to question 5. It is not accurate to call engineering the application of science, primarily because much of what engineers rely on is purely practical and empirical and, in some cases, without underlying scientific theory. Thus, engineering is still very much an art. This is not to deny that there is some application of science in engineering; it is simply to refute the notion that that is all engineering is.

2. A problem presented in a pure physical science course would involve an ideal case, such as the ideal transformer, that assumed away all complications (such as resistance, flux leakage, eddy currents) that becloud the answer to a problem involving the essential elements of the case.

The same problem presented in an applied science course would not assume away all these qualifying variables of an actual device, but would provide the student with values for all these variables.

The same problem presented to an engineer would give only the essential input, for example, the input voltage and the frequency of the input alternating current along with a specification of the desired output voltage and current. The engineer would then have to approximate a solution in the form of an actual transformer by experimenting with models until the optimum transformer is discovered for the problem at hand.

3. Aerospace engineers design systems that travel above the surface of the earth.

Chemical engineers design processes for the large scale transformation of materials by chemical processes.

Civil engineers design major structures and the means of constructing them.

Electrical engineers design the means by which electrical energy is created, transferred, and used.

Industrial engineers design systems for producing goods and services on a large scale.

Mechanical engineers design systems for the conversion of energy into useful forms.

4. Some nontechnical areas of knowledge important to engineers are economics, political science, psychology, sociology, business realities (such as costs and return on investment), general knowledge about the problems of the many people that engineers must work with and knowledge of how to deal with them, knowledge of values and ethics and of the many ways that engineering has impacts on society.

5. Science and engineering differ in the following ways.

The basic processes characteristic of each

Predominant day-to-day concerns of each

Primary end products of each

Knowledge employed by each

ISBN 0-471-01702-7